FOCUS ON

MIDDLE SCHOOL

GEOLOGY

Teacher's Manual

3rd Edition

Rebecca W. Keller, PhD

REAL SCIENCE 4 Kids

Real Science-4-Kids

Focus On Middle School Geology Teacher's Manual—3rd Edition
ISBN 978-1-941181-56-0

Published by Gravitas Publications Inc.
www.gravitaspublications.com
www.realscience4kids.com

GRAVITAS
PUBLICATIONS

A Note from the Author

This curriculum is designed to engage middle school level students in further exploration of the scientific discipline of geology. The *Focus On Middle School Geology Student Textbook—3rd Edition* and the accompanying *Laboratory Notebook* together provide students with basic science concepts needed for developing a solid framework for real science investigation into geology.

The experiments in the *Laboratory Notebook* allow students to expand on concepts presented in the *Student Textbook* and develop the skills needed for using the scientific method. This *Teacher's Manual* will help you guide students through the laboratory experiments.

There are several sections in each chapter of the *Laboratory Notebook*. The section called *Think About It* provides questions to help students develop critical thinking skills and spark their imagination. The *Experiment* section provides students with a framework to explore concepts presented in the *Student Textbook*. In the *Conclusions* section students draw conclusions from the observations they have made during the experiment. A section called *Why?* provides a short explanation of what students may or may not have observed. And finally, in each chapter an additional experiment is presented in *Just For Fun*.

The experiments take up to 1 hour. Materials needed for each experiment are listed on the following pages and also at the beginning of each experiment.

Enjoy!

Rebecca W. Keller, PhD

Materials at a Glance

Experiment 1	Experiment 3	Experiment 4	Experiment 5	Experiment 6
pencil, pen colored pencils small jar trowel or spoon **Optional** binoculars	known mineral samples: calcite, feldspar, quartz, hematite (from local rock and mineral store or order online) * several rocks from backyard or nearby copper penny steel nail streak plate (unglazed white ceramic tile) paper scissors marking pen tape vinegar (small amount) lemon juice (small amount) eyedropper or spoon	Students will select materials and use them to make a model of Earth's layers **Chocolate Lava Cake** butter: 113 grams (1/2 cup) semi-sweet chocolate chips: 133 ml (1/2 cup + 1 Tbsp.) 2 whole eggs 2 egg yolks powdered sugar: 192 ml (3/4 cup + 1 Tbsp.) flour: 94 ml (1/3 cup + 1 Tbsp.)	student-made brittle candy (materials in Foods column on next page) 1 jar smooth peanut butter (for students with allergies to peanuts, whipped cream can be substituted) 118 ml (1/2 cup) crushed graham crackers plate or second cookie sheet materials to make a model volcano — student's choice	internet access (a program that unzips files may be needed) **Optional** printer and paper colored pencils

Experiment 2				
pencil pen colored pencils compass small jar or container with a lid small items to place in jar (student selected) **Optional** garden trowel				

Experiment 7	Experiment 9	Experiment 10	Experiment 11	Experiment 12
Some suggestions for student chosen model making materials: modeling clay of different colors marble or steel ball ingredients to make various colored cakes materials for making paper maché Styrofoam balls	gravel, about 1-2 liters** sand, about 1-2 liters** dirt (soil), about 1-2 liters** pottery clay, about 1-2 liters** water (4) Styrofoam cups, about 355 ml (4) 16 oz. clear plastic cups, glasses, or other clear containers pencil marking pen measuring cups graduated cylinder, 100 ml large bowl scissors plastic wrap or plastic bags cardboard or other material to make a trough strong tape utility or X-Acto knife bucket and/or outdoor area **Optional** stopwatch or clock with a second hand small piece of screen or coarse cloth	field notebook, 1-2 (new or existing) pencil and colored pencils small backpack water bottle snacks binoculars (inexpensive ones are fine; small, lightweight ones are easier to carry) field guide to the birds book (for example, *The Young Birder's Guide to Birds of North America,* by Bill Thompson, III) **Optional** magnifying glass bird feeders and birdseed camera cellphone or tablet bird identification app (such as free app from Audubon Society, http://www.audubon.org/apps)	steel needle bar magnet piece of cork tape medium size bowl water compass small object of student's choice to use for treasure	pencil pen imagination **Optional** notebook

Experiment 8				
2 liter (2 quart) plastic bottle with cap warm water matches blank paper				

* A Mineral Scale of Hardness Set of Minerals is available from Home Science Tools: http://www.hometrainingtools.com

** Quantities may vary. See Experiment 9, Just For Fun. Materials can be found where art, aquarium, or building supplies are sold.

Materials
Quantities Needed for All Experiments

Equipment	Materials	Foods

backpack, small
binoculars (inexpensive ones are fine; small, lightweight ones are easier to carry)
book, field guide to the birds (for example, *The Young Birder's Guide to Birds of North America*, by Bill Thompson, III)
bowl, large
bowl, medium
bucket (and/or outdoor area)
compass
cookie sheet, 1-2, approx. 30x36 cm (12"x14")
cups (4), plastic, clear, 16 oz.; or clear glasses or other clear containers
eyedropper or spoon
graduated cylinder, 100 ml
jar, small
jar, small (or container) with a lid
knife, utility or X-Acto
magnet, bar
measuring cups
needle, steel
penny, copper
plastic bottle with cap, 2 liter (2 quart)
plate or second cookie sheet
saucepan, 2 liter (2 qt)
scissors
spatula, 2
streak plate (unglazed white ceramic tile)
thermometer, candy
trowel or spoon
water bottle
 	
Optional
 	
bird feeders
camera
cellphone or tablet
magnifying glass
printer
stopwatch or clock with a second hand
	cardboard or other material to make a trough
	clay, pottery, about 1-2 liters (1-2 qts.)**
	cork
	cups (4), Styrofoam, about 355 ml (12 oz.)
	dirt (soil), about 1-2 liters (1-2 qts.)**
	gravel, about 1-2 liters (1-2 qts.)**
	items, small, (student selected) to place in jar
	matches
	materials, student selected, for making a model volcano
	materials, suggested, student selected for making model of Earth's layers:
	   modeling clay of different colors
	   marble or steel ball
	   ingredients to make various colored cakes
	   materials for making paper maché Styrofoam balls
	mineral samples (known): calcite, feldspar, quartz, and hematite (from local rock and mineral store or order online) *
	nail, steel
	notebook (field), 1-2 (new or existing)
	paper
	pen
	pen, marking
	pencil
	pencils, colored
	plastic wrap or plastic bags
	rocks, several, from backyard or nearby
	sand, about 1-2 liters (1-2 qts.)**
	tape
	tape, strong, such as duct tape
	
	Optional
	
	birdseed
	notebook
	screen or coarse cloth, small piece
		graham crackers, crushed, 118 ml (1/2 c.)
		lemon juice (small amount)
		peanut butter, 1 jar smooth (for students with allergies to peanuts, whipped cream can be substituted)
		snacks
		vinegar (small amount)
		
		Chocolate Lava Cake (Exper. 4)
		  butter: 113 grams (1/2 cup)
		  semi-sweet chocolate chips: 133 ml (1/2 cup + 1 Tbsp.)
		  2 whole eggs
		  2 egg yolks
		  powdered sugar: 192 ml (3/4 cup + 1 Tbsp.)
		  flour: 94 ml (1/3 cup + 1 Tbsp.)
		
		Brittle Candy (Exper. 5)
		  white sugar: 237 ml (1 cup)
		  light corn syrup: 118 ml (1/2 cup)
		  salt: 1.25 ml (1/4 teaspoon)
		  water: 59 ml (1/4 cup)
		  butter: 28 grams (2 Tbsp)
		  baking soda: 5 ml (1 teaspoon)

Other

internet access (a program that unzips files may be needed)

Optional

bird identification app (such as free app from Audubon Society, http://www.audubon.org/apps)

* A Mineral Scale of Hardness Set of Minerals is available from Home Science Tools: http://www.hometrainingtools.com
** Quantities may vary. See Experiment 9, Just For Fun. Materials can be found where art, aquarium, or building supplies are sold.

Contents

Experiment 1

Observing Your World

Materials Needed

- pencil, pen, colored pencils
- small jar
- trowel or spoon

Optional

- binoculars

Objectives

In this experiment students will collect data from their observations of the world around them and organize the data in a chart.

The objectives of this lesson are for students to:

- Make careful observations.
- Record and organize data.

Experiment

I. Think About It

Read this section of the *Laboratory Notebook* with your students.

Ask questions such as the following to guide open inquiry.

- *Do you think if you go outside and look at things carefully you will observe details that you hadn't noticed before? Why or why not?*

- *If you are riding your bike or walking outside, what kinds of things do you notice that tell you where you are?*

- *Do you think buildings are different in the city than they are in the country? Why or why not?*

- *Do you think there are the same kinds of landscapes in a city and in the country? Why or why not?*

- *What kinds of weather do you have in your area?*

- *Does weather affect the landscape where you live? Why or why not?*

II. Experiment 1: Observing Your World

In this experiment students will explore their local world—the world around them.

Have the students read the entire experiment, noting the experimental steps to be followed.

Objective: Have the students write an objective. Some examples:

- *To learn more about the place where I live.*

- *To better understand my hometown (or city, etc.)*

- *To observe my surroundings and find out more about Earth through my observations.*

Hypothesis: Have the students write a hypothesis. Have them think about what they might learn by doing this experiment. Some examples:

- *By making observations, I will better understand where I live.*

- *By making observations, I can understand how the area where I live changes over time.*

- *I will be able to assemble a list of features that describe where I live.*

❶-❸ Students walk to a place where they are standing on some type of ground (e.g., dirt, grass, concrete) and then observe what is around them. In the boxes provided, they will write or draw what they see. If binoculars are available, students can use them to observe the details of landforms and other features that are farther away.

❹-❺ Have students go to a place where they can dig up a small sample of dirt with a trowel or spoon. They can put the dirt sample in a small jar. Have them examine the dirt carefully, looking for color, texture, the presence of organic matter, and any other features they can observe. Have them record their observations.

❻ Have the students observe any man-made structures nearby. These can include things such as homes, roads, buildings, parks, fences, and utility poles.

❼ Have the students explore how dynamic features near their home may have changed the way the area looks. For example, heavy rains or snowmelt runoff might have caused erosion. Lightning might have damaged a tree.

❽ Have the students explore the history of the place where they live. Have them research when their house was built, when their city was founded, or when the land surrounding them became farms or industrial parks.

Results

Have the students use the chart to assemble the information they've collected.

III. Conclusions

Have the students review the results they recorded for the experiment. Have them draw conclusions based on the data they collected.

IV. Why?

Read this section of the *Laboratory Notebook* with your students.
Discuss any questions that might come up.

V. Just For Fun

Have the students imagine they got an email from someone on the planet Alpha Centauri Bb asking for information about what Earth is like. Have the students use the data they have collected to write a narrative describing the area where they live. Have them think about how they would describe the area so that someone who has not seen it could imagine it.

Experiment 2

Hidden Treasure

Materials Needed

- pencil, pen, colored pencils
- compass
- a small jar or container with a lid
- small items to place in jar (student selected treasure)
- garden trowel (optional)

Objectives

In this experiment students will explore some tools geologists use to study Earth.

The objectives of this lesson are for students to:

- Explore mapmaking.
- Use a compass to find directions.

Experiment

I. Think About It

Read this section of the *Laboratory Notebook* with your students.

Ask questions such as the following to guide open inquiry.

- *What different kinds of maps have you used? What did you use them for? What else do you think maps could be used for?*

- *What parts of making a map do you think would be easy? Difficult? Why?*

- *Why do you think geologists need to use maps?*

- *What problems could occur if you were using a map that was not accurate? Why?*

- *When do you think it would be helpful to use a compass? A GPS? Why?*

- *What advantages do modern geological tools have over older tools?*

- *What disadvantages might modern tools have?*
  (For example, the need for electricity or gas power, fragility, need for software updates, expense, etc.)

II. Experiment 2: Hidden Treasure

In this experiment students will explore mapmaking by creating a map for finding a hidden treasure. Students will then test the accuracy of their map by having a friend try to find the treasure that is hidden or buried.

Have the students read the entire experiment before writing an objective and a hypothesis.

Objective: Have the students write an objective. Some examples:

- *To explore mapmaking by creating a real map.*

- *To understand how maps are made and the difficulties that come up.*

- *To create and test a homemade map.*

Hypothesis: Have the students write a hypothesis. Some examples:

- *I can create a map that is 80-100% accurate and I will know this by how quickly my friend finds the treasure.*

- *By creating a map and having it tested, I will learn about the difficulties of mapmaking.*

- *I can test the accuracy of my map by having a friend find my buried treasure.*

EXPERIMENT

In this experiment students will make a map and use a compass to determine north, south, east, and west.

❶ Have the students gather some small objects to use as treasure and place them in a jar.

❷ The students will be making their own map to use in finding a treasure they will hide. Help the students select an area to map—the backyard, a park, or other open space.

❸-❹ Students are to draw the outline of their map in the space provided. They will be using their footsteps as a tool for making measurements to put on their map. Have them count their steps while walking heel-to-toe around a given space. This space can be square or rectangular or oddly shaped. Have them draw on the map the actual shape of each side of the space and note the measurements and any obstacles they may encounter as they measure the space.

❺ Help the students understand how to use a compass.

Using a Compass

The needle on a compass will always point north (N). To read a compass, line the needle up with the north (N) label. If you wanted to travel north, you would head in this direction. To go south, you would travel in the opposite direction.

Once you have determined which way is north, you can find any other direction in which you want to go. For instance, if you want to go northeast, you would line the compass needle up with the north label and then travel in the direction of the northeast (NE) label on the compass.

Indicating Directions on the Map

Students will note the four directions on their map, indicating north with "N" along with an arrow. Then they will mark south with an "S", east "E," and west "W," again drawing arrows.

It may be helpful to the students to have them place their map on the ground in the same orientation as the landscape features. Have them hold the compass and stand next to the map. Direct them to turn until the compass needle points to north (N) and then draw an arrow and an N on their map to indicate which way is north. North will probably not be at the top of their map.

Explain that south is opposite north and have them mark south on their map with an arrow and an S. Next explain that when facing north, east will be to the right and west to the left. Have them mark east and west on their map.

You can also have the students use the compass to determine the direction each side of their map is facing.

❻ Have the students add details to their map, indicating objects and distances between them.

❼ Have the students pick a location to bury or hide their treasure and indicate this on their map.

❽ Students are to give their map to a friend and see if the friend can use it to find the treasure.

Results

Have the students record the number of attempts their friend makes in order to locate the buried treasure. Also have the students note any help they have to give to the friend in finding the treasure. This help can be used to make adjustments to the map to make it more accurate.

For example:

- *The number of steps was counted incorrectly.*
- *Direction of travel was followed incorrectly.*
- *An object noted on the map was moved.*

III. Conclusions

Have the students review the results they recorded for the experiment. Have them draw conclusions based on the data they collected.

IV. Why?

Read this section of the *Laboratory Notebook* with your students.
Discuss any questions that might come up.

V. Just For Fun

Have the students evaluate their map. Ask questions to help with their evaluation. For example:

- *Was the map accurate enough for your friend to find the hidden treasure? Why or why not?*

- *What (if any) modifications to the map did you make? Can you think of other ways to improve your map?*

- *How might you make your map more accurate?*

- *Does the size of your feet compared with the size of your friend's feet make a difference?*

Have the students review the results of their experiment. Encourage them to think of ways the map could be improved and then revise the map, including a new location for the hidden treasure. Have them give the map to a friend and see how well the revised map works compared to the original map.

Have the students evaluate how well the revised map worked compared to the original version. Have them record their results in the space provided.

Experiment 3

Mineral Properties

Materials Needed

- known mineral samples:
  calcite
  feldspar
  quartz
  hematite
- several rocks from backyard or nearby
- copper penny
- steel nail
- streak plate (unglazed white ceramic tile)
- paper
- scissors
- marking pen
- tape
- vinegar (small amount)
- lemon juice (small amount)
- eyedropper or spoon

You can find minerals at a local rock and mineral store or order them online.

A Mineral Scale of Hardness Set of minerals is available from Home Science Tools
http://www.hometrainingtools.com

Objectives

In this experiment students will begin to explore the Earth by learning about minerals and rocks.

The objectives of this lesson are for students to:

- Use scientific processes for identifying minerals.
- Use data from charts and make their own charts.

Experiment

I. Think About It

Read this section of the *Laboratory Notebook* with your students.

Ask questions such as the following to guide open inquiry.

- *Do you think rocks stay the same or change over time? Why?*
- *Do you think all rocks are made the same way? Why or why not?*
- *Do you think dirt (soil) has anything to do with rocks? Why or why not?*
- *How do you think soil is made?*
- *What do you think rocks and minerals can be used for?*
- *Do you think there are ways to tell what kind of rock you have? Why or why not?*

II. Experiment 3: Mineral Properties

Have the students read the entire experiment before beginning.

Objective: Have the students write an objective. Some examples:

- *To determine what minerals are in my backyard.*
- *To test rocks for the minerals calcite, quartz, feldspar, and hematite.*
- *To learn how to use field tests to determine which minerals are in rocks.*

Hypothesis: Have the students write a hypothesis. The hypothesis should be specific and can be about their backyard rocks or the known minerals they test. Some examples:

- *Rocks in my backyard contain quartz.*

- *Rocks in my backyard do not contain quartz.*

- *Feldspar is harder than calcite.*

- *Quartz is softer than hematite.*

Experiment—Part I

▶ Using the Mohs hardness scale provided in the *Laboratory Notebook,* have the students test each of the four known minerals for hardness.

The way to use the scale is as follows:

- *If a fingernail is able to scratch the mineral, the hardness is below 2.5.*

- *If a fingernail is not able to scratch the mineral, but a copper penny does, then the hardness is between 2.5 and 3.*

- *If a copper penny cannot scratch the mineral but a steel nail can, then the hardness is between 3 and 5.5.*

- *If a steel nail cannot scratch the mineral but an unglazed ceramic plate can (the mineral leaves a "streak"), then the hardness is between 5.5. and 6.5.*

- *If the mineral is not scratched by the ceramic plate (does not leave a streak) then the mineral is harder than 6.5.*

Have the students record other observations, such as size, color, and texture of the mineral samples. Also, have the students note any crystalline properties. It is important for students to record what they actually observe even if it's not what they think the "right" answer would be.

Typical hardness values are:

- Calcite: 3
- Quartz: 7
- Feldspar: 6
- Hematite: 5-6

Experiment—Part II

▶ Using the ceramic streak plate, have the students record the color streak left by each known mineral. To determine the color streak, they will take each mineral and scrape it firmly across the streak plate.

The color left on the plate is the "streak color" of the mineral. The streak color can be different from the color of the mineral sample as a whole. If this occurs, have the students note any differences. Some minerals, such as calcite, have a white streak.

Have students record their results.

Experiment—Part III

❶ Have the students go to the backyard or a place nearby and collect several rocks that look different from each other in color and texture.

❷-❹ Have the students label with numbers the rocks they collected and then record the numbers in the chart provided.

Students will do a hardness test and a streak test for each rock. Have them record their results and then compare them to the data collected in Part I and Part II to see if they can determine whether any of the known minerals they tested can be found in the locally collected rocks. They may or may not be able to determine whether the minerals they tested are in the unknown rock samples.

Other minerals can be tested if desired.

III. Conclusions

Have the students review the results they recorded for the experiment. Have them draw conclusions based on the data they collected.

The second part of this section asks the students to consider the subjective aspects of the hardness test and the streak test. They are subjective in that one person might see a slightly different color or observe a different hardness for the same mineral than another person, or objects used for testing might vary in actual hardness. Have the students think about how the subjectivity of the test might change the results.

An objective test is one where the observers' opinions may matter less, though this is not always the case. Have the students discuss whether or not chemical analysis might be more objective and less subjective than either the hardness test or streak test.

IV. Why?

Read this section of the *Laboratory Notebook* with your students.
Discuss any questions that might come up.

V. Just For Fun

Students will test their rock and mineral samples first with vinegar and then with lemon juice to see if any of them will have a chemical reaction with these acids. Have them rinse the vinegar off the rocks before testing with lemon juice. Have them record their results.

Have the students compare their results to see if the rock samples reacted in the same way to lemon juice as they did to vinegar.

Students can look on the internet or at the library to find out how the known mineral samples would be expected to react with acid. If their results varied, discuss how acids can be different strengths and the strength of the acid could affect the outcome of their experiment.

Experiment 4

Model Earth

Materials Needed

- Students will select materials and use them to make a model of Earth's layers

Chocolate Lava Cake

- butter: 113 grams (1/2 cup)
- semi-sweet chocolate chips: 133 ml (1/2 cup + 1 Tbsp.)
- 2 whole eggs
- 2 egg yolks
- powdered sugar: 192 ml (3/4 cup + 1 Tbsp.)
- flour: 94 ml (1/3 cup + 1 Tbsp.)

Objectives

In this experiment students will explore Earth's layers by making a model of Earth.

The objectives of this lesson are for students to:

- Explore model making as a way to understand Earth's layers.
- Learn about the benefits and limitations of using models in scientific discovery.

Experiment

I. Think About It

Read this section of the *Laboratory Notebook* with your students.

Ask questions such as the following to guide open inquiry.

- *How far below the surface of the Earth do you think rocks, minerals, and dirt extend? Why?*

- *What else do you think is below the Earth's surface? Gas? Melted rocks? Water? Living things?*

- *What do you think is at the center of the Earth? Is the center hard? Soft? Both? Neither?*

- *Do you think it's important that Earth has layers? Why or why not?*

II. Experiment 4: Model Earth

In this experiment students will create a model of Earth using what they have learned in this chapter of the *Student Textbook*. Students will design a model based on their own ideas about what materials could be used to represent Earth's different layers.

Have the students read the entire experiment.

Objective: Have the students write an objective. Some examples:

- *To explore building a model of Earth.*

- *To understand how to create accurate models of Earth.*

- *To explore the benefits and limitations of building models.*

Hypothesis: Have the students write a hypothesis. Some examples:

- *I can create a model of Earth that accurately represents Earth's layers.*

- *I cannot create a model of Earth that accurately represents Earth's layers.*

- *I can create a model of Earth that teaches me something about Earth's layers.*

❶ Have the students use the *Student Textbook,* the internet, and/or the library to collect information about Earth's layers. The more research they do, the more accurate the model they will be able to build and the more they will learn about Earth's layers.

In the chart provided have the students list the features and depths of the layers.

❷ Have students think about the characteristics of each layer and imagine materials they could use to create the layer so that the characteristics match.

Students can use clay, food items, Styrofoam, cloth, felt, dirt, or any other materials they think would work for assembling a model of Earth.

❸ Help students think about which layers and features they want to represent in their model. Encourage them to represent as many layers as possible. Have the students choose the model materials from their chart.

For example:

- *Students might want to build a model that represents the roundness of the Earth. In this case modeling clay of different colors could be used because it is easy to mold. A rock or small Styrofoam ball could be used for the solid inner core, a different color clay for each layer, and cloth for the crust. The clay layers will all be the same consistency.*

- *They might want to represent the rigid lithosphere and the soft asthenosphere. In this case a pan with layered food items could be used (For example, peanut brittle (lithosphere) on top of peanut butter (asthenosphere). Some hard candy or a chocolate bar might represent the inner core.*

- *If they chose to use a Styrofoam ball, then there would be no layers.*

❹ Have the students assemble their model.

Results

❶ Have the students determine how well their model represents Earth's geology. The models are not going to be perfect, and in the process of model building some features will be more accurately represented than others.

❷ Have the students fill in the chart provided. There are no right answers and their answers will be based on what they actually observed.

III. Conclusions

Have the students review the results they recorded and evaluate their models. Help them explore how challenging model building can be and the difficulties that arise when trying to build accurate models.

IV. Why?

Read this section of the *Laboratory Notebook* with your students.
Discuss any questions that might come up.

V. Just For Fun

A fun way to model the Earth is to make a chocolate lava cake to represent the hot liquid of the Earth's outer core surrounded by the firmer mantle. Students can poke a walnut or cherry into the cake to represent Earth's inner core. They might add Earth's crust by sprinkling nuts over the top of the cake.

CHOCOLATE LAVA CAKE
butter   113 grams (1/2 cup)
semi-sweet chocolate chips   133 ml (1/2 cup + 1 Tbsp.)
2 whole eggs
2 egg yolks
powdered sugar   192 ml (3/4 cup + 1 Tbsp.)
flour   94 ml (1/3 cup + 1 Tbsp.)

Microwave butter briefly until melted. Stir in chocolate chips until melted. Mix in whole eggs and yolks, then powdered sugar. Stir in flour. Pour into custard cups thoroughly greased with butter. Bake at 190° C (375° F) until edges are set and centers are still soft, about 10-13 minutes. Be careful not to overbake.
Makes about 4.

Have the students evaluate their lava cake model and compare it to their first model.

Experiment 5

Dynamic Earth

Materials Needed

- student-made brittle candy (see below)
- 1 jar smooth peanut butter (for students with allergies to peanuts, whipped cream can be substituted)
- 118 ml (1/2 cup) crushed graham crackers
- plate or second cookie sheet
- materials to make a model volcano—student's choice

Materials for making brittle candy

Ingredients

237 ml (1 cup) white sugar
118 ml (1/2 cup) light corn syrup
1.25 ml (1/4 teaspoon) salt
59 ml (1/4 cup) water
28 grams (2 Tbsp) butter, softened
5 ml (1 teaspoon) baking soda

Equipment

2 liter (2 qt) saucepan
candy thermometer
cookie sheet, approx.
   30x36 cm (12x14 inches)
2 spatulas

Objectives

In this experiment students will build models to explore Earth's dynamics.

The objectives of this lesson are for students to:

- Explore model making for geological study.
- Observe how different layers of the Earth work together.

Experiment

I. Think About It

Read this section of the *Laboratory Notebook* with your students.

Ask questions such as the following to guide open inquiry.

- *How do you think mountains are formed?*
- *What do you think causes earthquakes?*
- *What do you think causes volcanoes?*
- *What happens to the surrounding area when a volcano erupts?*
- *What causes tectonic plates to move?*

II. Experiment 5: Dynamic Earth

Students will build a model using food items to examine how the movement of Earth's tectonic plates causes earthquakes to occur and mountains to form.

Have the students read the entire experiment.

Objective: Have the students write an objective. Some examples:

- *To explore how Earth's plates move on the putty-like asthenosphere.*
- *To experiment with a model of the crust, lithosphere, and asthenosphere to learn about plate tectonics.*
- *To explore the benefits and limitations of building models.*

Hypothesis: Have the students write a hypothesis. Some examples:

- *I can create a model that will help me understand how plate tectonics works.*

- *Making a model will show me something I did not expect about plate tectonics.*

❶ Students will make brittle candy that will represent tectonic plates in their model.

Help the students follow the recipe to make the brittle candy. It can be tricky to determine when the candy has cooked long enough. Make sure the students stir and check the candy often, noting the temperature and using fresh cold water each time they test the candy.

❷-❺ Once the candy has cooled, have the students create a model of the three layers—asthenosphere, lithosphere, and crust—using peanut butter, brittle candy, and the peanut butter/graham cracker mixture.

If a student has allergies to peanuts, whipped cream can be used in place of peanut butter.

❻ In the chart provided, have the students write the layer of Earth that each food item represents.

❼ Have the students move the brittle candy pieces on top of the peanut butter.

Have them bump the pieces together, scrape them alongside each other, and move the pieces up and down with respect to each other.

The students should pay special attention to the graham cracker/peanut butter topping. This layer represents the crust and is where the most visible changes will occur.

Have the students explore what happens when the brittle candy pieces are moved slowly, quickly, and collide or move past each other.

Explore with the students how well this works as a model of plate tectonics.

❽ Have the students write their observations in the box provided.

Results

Have the students review their observations and use them in recording their answers to the questions in the chart provided.

Have the students discuss the ways in which their model reflects the theory of plate tectonics and how it does not. For example:

- *The textures of the three foods may adequately represent the differences between the crust, lithosphere, and asthenosphere.*

- *The peanut butter may be pushed up to the surface where the "crust" is. This does not appear to happen in reality.*

- *The graham cracker/peanut butter crust may slide too easily off the brittle candy pieces, which doesn't reflect reality.*

- *The graham cracker/peanut butter crust mounds up like mountains.*

III. Conclusions

Have the students review the observations and results they recorded for the experiment and use them to evaluate their model.

Help the students explore how challenging model building can be and the difficulties that arise when trying to build accurate models.

IV. Why?

Read this section of the *Laboratory Notebook* with your students.
Discuss any questions that might come up.

V. Just For Fun

Have the students do online or library research to find a model volcano they can build. Once they have decided which model they want to make, they will create their own experiment, following the format of previous experiments.

Experiment 6

Using Satellite Images

Materials Needed

- computer with
  internet access
  (a program that unzips files
  may be needed)

Optional

- printer and paper
- colored pencils

Objectives

In this experiment students will explore satellite images and how they are used by geologists.

The objectives of this lesson are for students to:

- Learn how to use online resources to collect data.
- Observe how technology such as satellite imagery can be used to study changes in Earth's system.

Experiment

I. Think About It

Read this section of the *Laboratory Notebook* with your students.

Ask questions such as the following to guide open inquiry.

- *What do you think might happen to a beach during a hurricane? Which of Earth's spheres do you think could be affected by a hurricane?*

- *What do you think might happen to trees during a forest fire? Which of Earth's spheres do you think could be affected by a forest fire?*

- *What do you think might happen to rivers and lakes during a drought? Which of Earth's spheres do you think could be affected by a drought?*

- *How many of Earth's spheres do you think can be affected when a volcano erupts? Which ones? How?*

II. Experiment 6: Using Satellite Images

Have the students read the entire experiment.

Objective: Have the students think of an objective for this experiment (What will they be learning?).

Hypothesis: There is no hypothesis for this experiment since it is an observational experiment.

EXPERIMENT

❶ Have the students go to the US Geological Survey (USGS) website and spend some time looking through the collections of sets of satellite images that show changes to Earth's surface over time. As of this writing the URL for the images is:

https://remotesensing.usgs.gov/gallery/

❷-❺ Students will be looking for images that they think show changes that have affected each of the different spheres of Earth. Have them select one set of images for each of the spheres: the geosphere, the atmosphere, the hydrosphere, and the biosphere.

Under each set of images on the USGS website there will be a *Download Image* section. Have the students click on the small size image file and download it. If they get a "zip file" that contains the images, they will need a program that unzips files for them to be able to look at the images. Have them make a folder on the computer and put the downloaded image files in it.

If possible, have them print the images, label them, and insert them in their *Laboratory Notebook*. If a printer isn't available or the images don't print well enough to be usable, students can refer to the files on the computer as needed.

Observations will be noted in the *Results* section.

Results

For each sphere, have the students fill in the information requested in the charts. They are asked to make a rough sketch of the images they are referencing even though they may have printed them out. Making sketches will help the students look more carefully at the images. The sketches do not need to be highly detailed and can be done in color or black and white. They can be done on separate pieces of paper if students want a larger area for drawing.

Note: Websites change from time to time. Should the referenced USGS website become unavailable, check to see if the Landsat images are now in a different location on the USGS site. If you can't find them, students can use one of the websites in the *Just For Fun* section. In this case, they probably won't be seeing side-by-side before and after images, but they should still be able to find images where changes have occurred due to storms, volcanic eruptions, etc.

Another possible resource for satellite imagery is the European Space Agency (ESA) site: http://www.esa.int/Our_Activities/Observing_the_Earth/

III. Conclusions

Have the students discuss what they learned by observing images of Earth taken from space and how they think satellites have changed what we know and can discover about Earth.

IV. Why?

Read this section of the *Laboratory Notebook* with your students. Discuss with your students how satellite images can help scientists observe Earth's interconnected systems and how natural and human activities interact with these systems. Discuss any questions that might come up.

V. Just For Fun

Students will spend some time viewing more satellite images on two NASA websites—or you can have them choose one of the sites. Have the students follow the directions in the *Laboratory Notebook* for using each website. The Earth Observatory website is the more straightforward of the two.

NASA's Earth Observatory website has photographs and videos taken from the International Space Station.

http://earthobservatory.nasa.gov/

Directions provided to students: On the top menu bar click on *Images* to find collections by topic. Select *Natural Hazards* and explore satellite images of several natural disasters. Go back to the Home pages and scroll down to find the *Special Collections* groups of photographs to explore.

NASA's Gateway to Astronaut Photography of Earth

http://eol.jsc.nasa.gov/Collections/EarthFromSpace/

Directions provided to students: Select a topic from the left menu bar. On the next screen select what you would like to view and click *Start Search*. Play around to see what else you can find on this site.

Have the students answer the questions about their observations. Have them discuss the new things they discovered.

Note: Websites change from time to time. If the above referenced websites work differently, help the students navigate to the information they are looking for. Should the above referenced websites become unavailable, browser searches can be done to find other NASA or ESA websites that have satellite images of Earth.

Experiment 7

Modeling Shaky Ground

Materials Needed

- Jell-O or other gelatin (any color/any flavor)
- graham crackers
- marshmallows (Optional: large and small)
- toothpicks
- baking pan—24 cm x 28 cm (9.5" x 11")
- student-selected materials (Just For Fun)

Objectives

In this experiment students will learn more about the layers of the geosphere and earthquakes through model making.

The objectives of this lesson are for students to:

- Explore the features of the geosphere by making a model.
- Explore earthquakes.

Experiment

I. Think About It

Read this section of the *Laboratory Notebook* with your students.

Ask questions such as the following to guide open inquiry.

- *Which layers of Earth do you think are involved in earthquakes?*

- *What do you think makes an earthquake happen?*

- *Do you think an earthquake can happen anywhere on Earth? Why or why not?*

- *How do you think people could prepare for earthquakes?*

- *How do you think it would be helpful if scientists could predict when and where an earthquake will occur?*

II. Experiment 7: Modeling Shaky Ground

Have the students read the entire experiment.

Objective: Have the students think of an objective for this experiment (What will they be learning?).

Hypothesis: Have the students think of a hypothesis for this experiment.

EXPERIMENT

❶ Help the students make gelatin according to the instructions on the box. Have them pour the gelatin into the baking pan and refrigerate it.

❷ When the gelatin has set completely, have the students remove it from the refrigerator. Have them place graham crackers on top of the gelatin. Have them decide how they want to divide the crackers. They can make all the pieces the same size or different sizes. Help the students observe how the crackers are arranged. (Are they touching? Far apart? Flat? At an angle? What else can they notice?) Have them draw and label their observations in the box provided. They can draw a top view, a side view, or both.

Example of basic diagram of experimental setup:

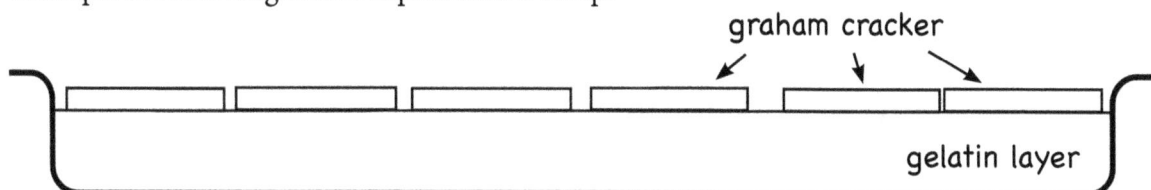

graham cracker

gelatin layer

❸ Students are directed to press down on the graham crackers at one edge of the pan and observe what happens. (For example: How far down does the graham cracker move? When they press on one graham cracker, do the others move? How? What else can they observe?)

Pressing on a graham cracker is expected to cause movement in the Jell-O.

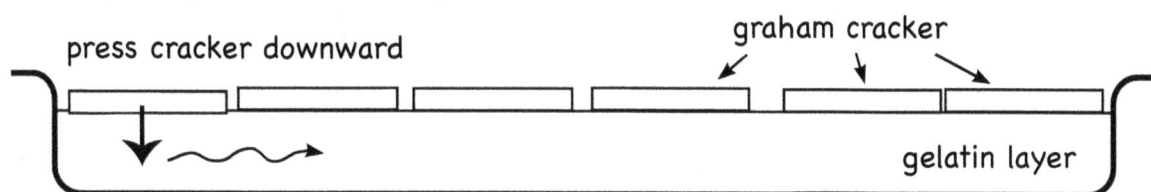

press cracker downward

graham cracker

gelatin layer

❹ Have students repeat as many times as they'd like, experimenting with pressing on different crackers, more than one at a time, using more or less pressure, pulsing the crackers, etc.

❺ In the two boxes provided, have them draw and label their observations. They are invited to use additional sheets of paper for more exploration.

❻ The students are directed to place marshmallows on the graham crackers. Have them experiment with pressing on different crackers, more than one at a time, using more or less pressure, pulsing the crackers, etc. What happens to the marshmallows?

Have them record their results in the *Results* section.

❼ Have the students try stacking the marshmallows, repeat Step ❻, and observe what happens.

Have them record their results in the *Results* section.

❽ Have them use toothpicks to secure the marshmallows to the graham crackers, then repeat Step ❻. What happens? Can they get the marshmallows to move?

Have them record their results in the *Results* section.

III. Conclusions

Have the students think about what they learned by building a model of Earth's layers. Help them think about what the gelatin and graham crackers represent. What does the model as a whole represent? What did they learn about earthquakes by building this model? What are some limitations of this model and models in general?

IV. Why?

Read this section of the *Laboratory Notebook* with your students.
Discuss any questions that might come up.

V. Just For Fun

Students are directed to create their own experiment based on the one they just performed and using the same format.

Help the students think about how they might change this experiment to learn more about earthquakes. For example, what would happen if, instead of using marshmallows, they built different types of structures from different materials? Could they use something besides graham crackers? What would happen if they made gelatin of different thicknesses by adding more or less water when mixed? Have them think about other changes they could make.

Have the students use separate paper to write up their experiment.

Experiment 8

Exploring Cloud Formation

Materials Needed

- 2 liter (2 quart) plastic bottle with cap
- warm water
- matches
- blank paper

Objectives

In this experiment students will explore how clouds are formed.

The objectives of this lesson are for students to:

- Understand that different factors in the atmosphere affect cloud formation.
- Use simple tools for exploration.

Experiment

I. Think About It

Read this section of the *Laboratory Notebook* with your students.

Ask questions such as the following to guide open inquiry.

- *What do you think Earth would be like if it did not have an atmosphere?*

- *Do you think life could survive on Earth if the atmosphere did not carry water? Why or why not?*

- *Do you think the atmosphere is involved in the creation of rain and snow? Why or why not?*

- *Why do you think atmospheric pressure is important to life?*

- *What do you think would happen to the atmosphere if Earth's gravity were weaker? Would this affect clouds? Why?*

II. Experiment 8: Exploring Cloud Formation

Have the students read the entire experiment.

Objective: Have the students write an objective. Some examples:

- *To make clouds in a bottle.*

- *To understand how clouds are made.*

- *To see what happens to air and water in a bottle if a lit match is thrown in.*

Hypothesis: Have the students write a hypothesis. Some examples:

- *The air in the bottle will make a cloud.*

- *The air in the bottle will change as the bottle is squeezed and released.*

- *The lit match will go out when it hits the water.*

EXPERIMENT

❶ Have the students pour warm water into the plastic bottle until it is about 1/4 full and then put the cap on the bottle. Putting the cap on the bottle will cause water vapor to form.

❷ Have the students light a match, remove the cap from the bottle, and then drop the match in the bottle and quickly replace the cap. Students may need help with this step since the bottle cap must be removed and replaced quickly in order to keep water vapor from escaping.

❸ Have the students squeeze the plastic bottle near the bottom and release while observing what happens to the air in the bottle. Squeezing the bottle will increase the air pressure inside. Releasing the bottle will cause the air pressure to decrease, lowering the air pressure and thus cooling the air as it expands. This should cause clouds to appear. If clouds don't appear, have the students try using warmer water.

❹ Have the students record their observations in the chart in the *Results* section. Have them write down as many details as possible.

❺ Have the students repeat this experiment, filling the plastic bottle 1/2 full, 2/3 full, and then almost full, emptying the bottle each time and starting with fresh warm water. Have them record their observations each time.

Results

A chart is provided for the students' observations.

III. Conclusions

Have the students review the results they recorded for the experiment. Have them draw conclusions based on the data they collected.

Based on the observations they have recorded, have the students write down whether they think any of the variables might have changed during the different steps of the experiment. Explain that *ratio* means the quantity of the different factors relative to each other.

Ask questions such as the following:

- *Was the water the same temperature for each part of the experiment?*
- *Was the cap on or off for a longer time?*
- *Was the bottle harder or easier to squeeze?*
- *Was there more or less air in the bottle?*
- *Was there more or less smoke?*

IV. Why?

Read this section of the *Laboratory Notebook* with your students.
Discuss any questions that might come up.

V. Just For Fun

Students are asked to observe the clouds daily for two weeks or more and record their observations of the clouds and the weather conditions. Have them include sketches of the clouds they see. A chart format is provided for them to follow as they make additional chart pages on blank paper.

Have them check online for the humidity, dew point, and low and high temperatures and record this data daily on their chart.

At the end of the experiment, have the students analyze the data they have recorded to look for relationships between weather conditions, types of clouds, temperature, humidity, and dew point. Have them record their conclusions and fasten their chart pages in their *Laboratory Notebook*.

Experiment 9

What Makes an Aquifer?

Materials Needed

- gravel, about 1-2 liters (1-2 qts.)*
- sand, about 1-2 liters (1-2 qts.)*
- dirt (soil), about 1-2 liters (1-2 qts.)*
- pottery clay, about 1-2 liters (1-2 qts.)*
- water
- (4) Styrofoam cups, about 355 ml (12 ounce) size
- (4) 16 oz. clear plastic cups, glasses, or other clear containers
- pencil (for making holes in Styrofoam)
- marking pen
- measuring cups
- graduated cylinder, 100 ml
- large bowl
- scissors
- plastic wrap or plastic bags
- cardboard or other material to make a trough
- strong tape such as duct tape
- utility knife or X-Acto knife
- bucket and/or outdoor area

Optional

- stopwatch or clock with a second hand
- small piece of screen or coarse cloth

* Quantities may vary. See *Just For Fun* section. Materials can be found where art, aquarium, or building supplies are sold.

Objectives

In this experiment students will explore permeability and porosity.

The objectives of this lesson are for students to:

- Observe how permeability and porosity affect how water travels through a substance.
- See that math can be used to quantify porosity.

Experiment

I. Think About It

Read this section of the *Laboratory Notebook* with your students.

Ask questions such as the following to guide open inquiry.

- *What do you think happens to rain when it falls on the ground?*
- *Do you think the type of ground that rain falls on makes a difference in what happens to the rainwater? Why or why not?*
- *After it rains, which do you think will dry out more quickly—sand or garden soil? Why?*
- *Do you think it can be important to know whether rain will soak into the ground in a certain area or whether it will run off? Why?*
- *Do you think knowing how water interacts with different materials could help with flood control, farming, or obtaining drinking water? Why?*
- *What do you think happens with water that is under the surface?*

II. Experiment 9: What Makes an Aquifer?

Students will experiment with permeability and porosity.

Have the students read the entire experiment.

Objective: Have the students write an objective (What will they be learning?). Some examples:

- *To see how fast water runs through different materials.*
- *To find out which materials will hold more water.*

Hypothesis: Have the students write a hypothesis. Some examples:

- *Water will go through each material at a different speed.*

- *Some materials will hold more water than others.*

EXPERIMENT

Part I

In this part of the experiment students will test several natural materials for permeability—the ability of a porous material to allow water to pass through it. Before they begin the experiment, you can have them guess the relative permeabilities of the different materials.

❶ Have the students use a pencil to poke holes in the bottom of each of four Styrofoam cups. Have them put the same number of holes in each cup and make all the holes about the same size.

❷ Have the students measure 250 ml (1 cup) of each material (sand, gravel, dirt, and clay), put each material in its own cup, and label each cup with the material it contains.

❸ Have them hold each cup over a large bowl and pour 120 ml (4 ounces) of water into the cup and note how long it takes for the water to drain through each material. They can use a timer with a second hand or visually note how quickly or slowly the water runs through each material compared to the others.

❹ In the *Results* section, have them note how long it takes the water to go through each material and also write down any other observations.

Results

Part II

In this part of the experiment students will test for and calculate porosity. Porosity describes the size and quantity of void, or empty, spaces between the particles of a material. The porosity of a material describes the quantity of liquid it can hold, and the permeability of the material describes the speed at which liquid can flow through the material. It is possible for a material to be porous but not permeable (impermeable) if the pores (void spaces) are not connected.

❶ Have the students take four 16 oz. or larger clear plastic cups, glasses, beakers, or other clear containers. Have them measure 350 ml (1.5 cup) of each material (sand, gravel, dirt, and clay), put it in its own container, and label each container with the material it contains.

❷ Students will measure 100 ml of water in the graduated cylinder. They then will take one of the containers and slowly pour water into the material in it, trying to avoid making a depression in the top surface of the material. Have them pour water into the container until the water is up to the top of the material. Have them note how much water is left in the graduated cylinder and then record their observations in the *Results* section.

❸ Repeat Step ❷ for the other three materials.

Results

Have the students use the chart in this section to record their results and calculate the porosity of each material.

❶ Students will calculate how much water was held in the pores of the first material that was tested. To get this number have them subtract the amount of water that was left in the graduated cylinder from the 100 ml of water that they started with.

❷ Have the students divide the amount of water they were able to pour into the material by the total amount of material and express the answer as a percentage.

For example: If they were able to add 90 ml of water to one of the materials, the calculation would be:

$$90 \text{ ml water} \div 350 \text{ ml material} = .2571 = 25.71\% \text{ porosity.}$$

❸ Have them repeat Steps ❶-❷ for the remaining three materials.

III. Conclusions

Have the students answer the questions. Have them review the results they recorded for the experiment and draw conclusions based on the data they collected. What did they learn about permeability and porosity? Do they think permeability and porosity are related? Why or why not? What other conclusions can they draw from their observations?

IV. Why?

Read this section of the *Laboratory Notebook* with your students.
Discuss any questions that might come up.

V. Just For Fun

Build aquifers!

Students will explore aquifers through model making. It's a fun experiment to do outdoors but can also be done inside.

❶ Have the students review the results of their permeability and porosity tests and think about how they could layer these materials to make aquifers with different flow rates.

❷ Have the students discuss how they might make a container to hold the aquifer materials. One possibility is to take cardboard and fold it into a U shaped trough. Help the students determine the size trough they want to construct and have them measure the cardboard accordingly. Help them use a utility knife or X-Acto knife to cut the cardboard. It can be helpful to score the cardboard along the fold lines, and duct tape can be used to reinforce the folds. Students can then cover the cardboard with plastic wrap or plastic bags. You could also take students to a hardware or building supply store to look for other materials that could be used to make the trough, or you may have other materials around the house.

❸ Have the student think about different types of aquifers and flow rates and how the materials they tested could be layered in the trough to make a certain type of aquifer, for example one that water passes through quickly, one that holds water in one place in a reservoir, and long or short aquifers.

❹ Have the students use fresh materials for each aquifer they build. Have them layer two or more materials in the trough and predict what will happen when water is added. It's helpful for the students to fasten a piece of screen or coarse cloth on the ends of the trough to keep the materials from washing out.

In the chart in the *Results* section, have the students list the materials and the order in which they are layered in each aquifer made.

❺ Once the aquifer is built, students can add water and can experiment with different ways of doing this. They can tilt the aquifer and slowly pour water on the top end, seeing what happens if the aquifer is titled at different angles. They can pour water over the surface of the aquifer as if it were rain and see how this works with the aquifer.

Have them record their observations in the *Results* section.

❻ Have the students repeat Steps ❹-❺ to make different aquifers. Have them think about what properties they would like the aquifer to have, how many layers, the type of material in each layer, how the order of the layers will affect the aquifer, etc. There may be other materials around the house that they could include, such as other types of soil, sand, gravel, or small rocks.

Results

Charts are provided for students to record their observations.

When students have finished the experiment, have them review their results for the different model aquifers they built and then write a summary of their data. Next, have them use the data in their summary to draw conclusions about what they learned about aquifers by doing this experiment including how they think their models compare to real aquifers and why. Space is provided for the summary and the conclusions.

Experiment 10

My Biome

Materials Needed

- field notebook, 1-2 (new or existing)
- pencil and colored pencils
- small backpack
- water bottle
- snacks
- binoculars (inexpensive ones are fine; small, lightweight ones are easier to carry)
- field guide to the birds book (for example, *The Young Birder's Guide to Birds of North America*, by Bill Thompson, III)

Optional

- magnifying glass
- bird feeders and birdseed
- camera
- cellphone or tablet
- bird identification app (such as free app from Audubon Society, http://www.audubon.org/apps)

Objectives

In this experiment students will use direct observation to learn about the biome in which they live.

The objectives of this lesson are for students to:

- Make and record observations.
- Observe the interrelationships of plants, animals, weather, and geological formations.

Experiment

I. Think About It

Read this section of the *Laboratory Notebook* with your students.

Ask questions such as the following to guide open inquiry.

- *What details about plants and animals do you think you would observe if you looked carefully while walking around outside?*

- *Do you think you would notice things you haven't observed before if you walked around outside slowly and paid careful attention to what's around you? Why or why not?*

- *Do you think different animals and plants would live on a mountain than on the plains? Why or why not?*

- *Do you think different plants and animals live in dry climates than in wet climates? Why or why not?*

- *Do you think it would have any effect on an ecosystem if all the plants of one kind died? Why or why not?*

- *Why to you think a certain type of plant can only live in a certain kind of soil? In a certain climate?*

- *How do you think Earth's different spheres work together in a biome? (biosphere, hydrosphere, atmosphere, geosphere)*

II. Experiment 10: My Biome

Have the students read the entire experiment.

Objective: Have the students write an objective (What will they be learning?). Some examples:

- *To observe the environment where I live.*

- *To look at things in nature and figure out what kind of biome I live in.*

- *To observe plants and animals in my area and record my observations.*

Hypothesis: Have the students write a hypothesis. Some examples:

- *I will be able to make enough observations of plants and animals and other things to tell what biome I live in.*

- *By observing the environment outdoors, I can tell what biome I live in.*

- *I live in a city so I won't be able to tell what biome I'm in.*

EXPERIMENT

❶ Have the students pack a small backpack with a water bottle, pencils, an existing or new field notebook, and a snack. They can also take a magnifying glass and binoculars.

❷ Students are to take a one to two hour hike in the surrounding environment. Have them think about what route they would like to take on their hike. If you live in a city, help them decide where they could hike to see the most plants and animals.

Encourage the students to look all around them—to observe living things both big and small, up high and down low. Have them take their time and observe carefully.

❸ Encourage the students to carefully observe plant and animal life, where the plants and animals live, and what they are doing. They are also asked to observe landforms, rocks, and soils, and note how these affect plant and animal life. If you think it would be of interest to them, you can suggest that they make a map of their hike and mark on it where they made their favorite observations.

❹ Students are to describe the types of plants and animals (both large and small) they see and observe their interactions. Do they see insects on plants? Are there dogs playing in the grass? Are cats looking for birds or insects to catch? Are birds eating worms? Are plants blooming or going to seed? Are animals and insects eating? What are they eating?

❺ Have the students observe the weather and how the weather affects the plant and animal life. Is there snow on the ground and do the plants look lifeless? Is it warm and sunny and are there insects out foraging for food? Is it raining? Or windy? Are the birds cheeping or sitting quietly waiting for a storm to pass? Have them record their observations.

❻ In the *Results* section, have the students summarize their observations.

Results

Have the students review their notes and select the observations they think are the most important in describing their biome. In the chart provided, have them summarize their observations.

III. Conclusions

Have the students review the observations they recorded for the experiment and use the chart in the *Student Textbook* to determine the type of biome they live in. Have them record the name of their biome, describe it based on their notes and the *Student Textbook* chart, and list any unique features they discovered. Thinking about how they would describe their biome to someone who has never visited the area might help students decide which features to include in their description.

IV. Why?

Read this section of the *Laboratory Notebook* with your students.
Discuss any questions that might come up.

V. Just For Fun

What's that Bird?

In this experiment, students get a basic introduction to bird watching—an inexpensive hobby that can be done anywhere. For many, it is a fascinating way to spend time outdoors and learn more about nature. Have them record their observations in their existing field notebook or start a new one specifically for birds. The following list contains suggestions for getting started, and you and your students can choose to try them all or try the ones that are most interesting.

- Get a good field guide to the birds book that has pictures and information about birds in your area. For the US, a suggested book is the *The Young Birder's Guide to Birds of North America*, by Bill Thompson, III. This book was written for kids and is less overwhelming for beginning birdwatchers. Another good filed guide is *Peterson Field Guide to Birds of North America* which is more extensive. There are also other good field guides available. Have the students spend some time looking at the pictures and have them note any pictures of birds they think they have seen. Have them look at birds outside and then use the field guide to try to identify them. Binoculars are helpful.

- If you have space and a place that is safe for the birds, bird feeders can be set up in your yard. Students can research bird feeders to find the best ones for your location, which birds are most likely to be in your area, and what type of seeds they are most likely to eat. If you have a store that specializes in wild birds, this is a great resource for information and supplies. Many stores that sell pet supplies will have some birdseed available. Students can tend the bird feeders and observe the different birds that come to eat. Have them note the species of birds, the time of day they arrive, and how they interact. For a long-term project, students can observe seasonal changes in bird populations.

• Encourage students to get outside and look for birds. Have them take their hiking gear (backpack, binoculars, field guide to the birds book, field notebook, pencils, water, and snack) and go for a walk anywhere outside. They can use their field guide book and binoculars to help identify birds and then make notes in their field notebook. Have them note the types of habitats where they see the most birds and that different birds live in different habitats.

• Pishing is a technique used by birders to attract birds. The birder stands still and makes a sound like "psshh, psshh, psshh." This can take some patience while waiting for the birds to come. Students will want to be in an area where there are various bird habitats and likely to be lots of birds.

• Apps are available for bird identification. The Audubon Society has a free one that has photos of many different birds and information about each. It also includes bird calls (http://www. audubon.org/apps). Another app is called *Peterson Birds — A Field Guide to Birds of North America* and is available for a small fee. With a little research, students will be able to find other apps as well. Note that some apps may work only where there is cellphone or wifi connection, and not all apps will work with all devices.

For students who are interested in bird watching, there are various websites available for further research. The Audubon Society website is a good resource for bird information (http://www. audubon.org/). Another site is called eBird (http://ebird.org/) and has a lot of information about birds, including maps of cities and towns showing the local birding "hotspots," which are locations where the greatest variety of bird species have been spotted in one place.

Experiment 11

Finding North

Materials Needed

- steel needle
- bar magnet
- piece of cork
- tape
- medium size bowl
- water
- compass
- small object of student's choice to use for treasure

Objectives

In this experiment students will explore magnetic force and Earth's magnetic field.

The objectives of this lesson are for students to:

- Make a simple tool to test for magnetic force.
- Observe how a magnetized needle will align to Earth's magnetic field.

Experiment

I. Think About It

Read this section of the *Laboratory Notebook* with your students.

Ask questions such as the following to guide open inquiry.

- *How do you think a compass works? What would you use it for?*
- *How do you think you could test for the geomagnetic field?*
- *How do you think Earth's magnetic field is created?*
- *Do you think is important that Earth has a geomagnetic field? Why or why not?*
- *How do you think the magnetosphere is created? Is the magnetosphere important? Why or why not?*

II. Experiment 11: Finding North

Have the students read the entire experiment.

Objective: Have the students write an objective (What will they be learning?). Some examples:

- *To make a compass that works.*
- *To find out how to find north.*
- *To see if a needle can become magnetized and used to find north.*

Hypothesis: Have the students write a hypothesis. Some examples:

- *A compass can be made with a needle and used to find north.*

- *A bar magnet will make a needle into a magnet.*

- *The direction of north can be found with a homemade compass.*

EXPERIMENT

Before the experiment begins, take a cork (such as from a wine bottle) and cut off a piece big enough to float upright in water and keep a needle taped to the top of it dry.

❶ To magnetize the needle, have the students take the bar magnet and slowly stroke the needle against it for about 45 seconds.

❷ Have the students test the needle to make sure it's magnetized before taping it to the cork. Have them find an object that a magnet will stick to and then see if the needle will stick to that object. If the needle doesn't stick, have them rub it against the bar magnet for a while more, and then test it again.

❸ Have them center the magnetized needle on the top surface of the piece of cork and tape it in place.

❹ Have them pour water into the bowl until it is almost full and carefully place the cork in the center of the bowl so it is floating, needle-side up. The needle should not be touching the side of the bowl.

❺ Have the students observe what happens. When the cork stops turning on the water, the needle should be pointing in a north/south direction. However, this does not determine which end of the needle is pointing north.

❻ Have the students write about and/or draw their observations in the box provided.

❼ In order to find out which end of the needle is pointing north, students are asked to make a map. In the box in the *Results* section, have them draw a simple map of the walls of the room where the experiment is being performed. They should indicate the doorway(s) and window(s) to help orient the map. They may also indicate furniture or other features if they wish to.

❽ Have the students place their map on the table next to the bowl of water containing the compass and then turn the map so that it is in the same orientation as the walls of the room. Now have them draw a line in the middle of the map that matches the orientation of the needle. This is the north/south direction.

❾-❿ Help the students think about the approximate location of sunrise and sunset in relation to the room they're in. Once they have noted the approximate location of east and west on their map, they will be able to tell which end of the needle is pointing north and can note this on their map. Since the locations of east and west are only an approximation on this map, they may not be at right angles (90°) from north.

Results

Space is provided for drawing the map.

III. Conclusions

Have the students review the results they recorded for the experiment and think about how easy or difficult it was to make a compass and find north. Have them draw conclusions based on the data they collected.

IV. Why?

Read this section of the *Laboratory Notebook* with your students.
Discuss any questions that might come up.

V. Just For Fun

Finding treasure!

In this experiment students will use a compass to make a map and then have a friend use the map and compass to find a hidden treasure.

❶ Have the students practice using the compass to find north, south, east, and west until they are comfortable with this. The following instructions are provided:

Practice finding North (N), South (S), East (E), and West (W) with your compass. The needle will always be pointing to North. Stand with the compass in front of you, flat in your hand and parallel to the ground. Hold the compass in the same position as you turn in different directions.

Turn your body until the needle lines up with N. You are now facing North.

Now turn to the right until your body is lined up with the E on the compass. You are now facing East and the needle will be pointing to your left.

Turn to the right again until you are lined up with the S on the compass. You are now facing South and the needle will be pointing in the opposite direction, behind you.

Turn again to your right until you are lined up with the W on the compass. You are now facing West and the needle will be pointing to your right.

North, South, East, and West are called the four cardinal directions. Each is 90° from the previous one, or 1/4 of a 360° circle. If your compass has degrees shown on its face, the directions will be: N 0° , E 90°, S 180°, and W 270°.

❷ This experiment can be done outdoors or indoors. Once students have decided where they want to have the treasure hunt, they are to draw a map of the outline of this area. A box is provided in the next section of the *Laboratory Workbook*. They are to find north and mark it on the map with an arrow in the proper orientation relative to the outline. If they wish, they can draw features such as trees on the map, but these won't be used as markers for finding the treasure.

❸ Have the students decide where they want to hide the treasure.

❹ Have them pick a starting point for the route to the treasure. They should put a marker at this location since some precision will be needed for the treasure to be found by a friend using the map. Have the students draw the starting location on their map.

❺ Students are to use the compass to chart a route to the treasure using the cardinal directions only (N, S, E, & W) and measuring distances by counting heel-to-toe steps. Have them draw and label each leg of the route including direction of travel and number of steps.

❻ Have the students give their map to a friend or family member to use to find the treasure. If the friend doesn't know how to use a compass, the students can teach them. Help the students observe any difficulties that may have occurred in using the map and think about how they could refine it. They can make more maps that increase in difficulty, and their friends can make maps for the students to try.

❼ Have the students record their observations in the space provided. Ask them questions such as the following:

- *What did you learn about using a compass?*

- *How easy or difficult was it to make and follow the map?*

- *Was your friend able to find the treasure? Why or why not?*

- *Did you have to give some hints? Why or why not?*

- *What changes could you make to your map to refine it?*

- *Did you make more maps that were different from the first one? Did they work better than the first one? Why or why not?*

- *Did you have any difficulty pacing off the length of the different legs of the route? Why or why not?*

❽ Have the students look at Experiment 2 and compare the map they just made with the one in that experiment. Have them record their comparisons of difficulty, accuracy, etc.

Experiment 12

Solve One Problem

Materials Needed

- pencil, pen
- imagination

Optional

- notebook

Objectives

In this experiment students will explore concepts of Earth system science.

The objectives of this lesson are for students to:

- Observe how a local event in one of Earth's spheres can affect the other spheres.
- Begin to understand that they can solve problems by analyzing the different factors involved.

Experiment

I. Think About It

Read this section of the *Laboratory Notebook* with your students.

Ask questions such as the following to guide open inquiry.

- *Do you think an erupting volcano would interact with more than one of Earth's spheres? How and which ones?*

- *Which spheres are involved when it rains? How might they be affected?*

- *How would an earthquake affect different spheres and which ones would it affect?*

- *Think about the area in which you live. How do the different spheres interact to make the area the way it is?*

- *How could a change in one sphere affect other spheres in the area where you live? How would this make your area different?*

- *What are some human activities that you think might affect various spheres in a negative way? Which spheres would be affected?*

- *What are some human activities that might affect various spheres in a positive way? Which spheres would be affected?*

- *Do you think scientists will someday be able to predict natural events ahead of time? Why or why not?*

II. Experiment 12: Solve One Problem

Have the students read the entire experiment.

Objective: Have the students write an objective (What will they be learning?). Some examples:

- *To find solutions to problems in the neighborhood.*

- *To see what problems the neighbors have.*

- *To find out how I can help my neighbors.*

Hypothesis: Have the students write a hypothesis. Some examples:

- *I can get my friends together to help a neighbor with a problem.*

- *I can make problem solving into a non-profit business so I can help lots of neighbors.*

- *I can identify several problems in the neighborhood.*

EXPERIMENT

In this experiment students are asked to observe their neighborhood and identify any problems they notice. They are then asked to think of solutions to these problems and to carry out a solution to one problem.

❶ Have the students spend some time walking through their neighborhood and noticing what people are doing and how they interact with each other, with animals and plants, etc. Have them take their *Laboratory Notebook* or a separate notebook with them so they can write down their observations, including details. Using a separate notebook will give them more space to make notes and they can then later summarize their observations in their *Laboratory Notebook*.

❷ In the space provided in the *Laboratory Notebook* have the students make a list of all the problems they find. Have them include any details they notice.

❸ Next, have them record which different spheres these problems influence. For example, trash in the yard interacts with: Earth's crust (trash in the dirt), the biosphere (animals eat the trash), the atmosphere (the trash gives off an odor), and possibly the hydrosphere (if it rains and the water becomes contaminated).

❹ Have the students examine the problems they have listed and imagine a solution to each problem that they could actually carry out. An important part of Earth system science is looking at human activities, seeing how they affect various spheres, and then finding solutions to problems that are created by these human activities.

Encourage the students to use their imagination in thinking of solutions to the problems they've noticed. Let them record their ideas even if they seem impractical. The important thing is for the students to collect data and begin to analyze them in a problem solving manner.

❺ Have the students explore how they might expand their solutions to solve the problems for a larger number of people. Here and in Step ❻ again let the students use their imagination even if their expanded solutions seem impractical.

❻ Have the students explore how they could turn one of their ideas into a business or non-profit organization that would allow them to generate resources and jobs for others while helping to solve more problems.

Results

A chart is provided with questions that will help students summarize their results. Have them answer the questions using the data they've collected in this experiment.

III. Conclusions

Have the students discuss what they have learned about the interaction of Earth's spheres and Earth system science. Then have them review the charts they made during this experiment and describe how understanding Earth system science — how Earth is put together and how it works — can help give them the power to solve real-life problems.

IV. Why?

Read this section of the *Laboratory Notebook* with your students.
Discuss any questions that might come up.

V. Just For Fun

Have the students review the notes they made about the problem they'd like to solve. They are asked to come up with one or more ways to promote their idea to solve the problem. Some suggestions given are: make a flyer to hand out, write a newspaper article or letter to the editor of a local newspaper, start a blog, contact a local radio or TV station about doing a news story, make an informational video about their project. The important idea here is for students to begin to look at different ways they can make a positive impact on their local community. They may not actually perform any of these tasks, but encourage them to think of as many ideas as possible even if they come up with something wild or improbable.

More REAL SCIENCE-4-KIDS Books
by Rebecca W. Keller, PhD

Building Blocks Series yearlong study program — each Student Textbook has accompanying Laboratory Notebook, Teacher's Manual, Lesson Plan, Study Notebook, Quizzes, and Graphics Package

Exploring the Building Blocks of Science Book K (Activity Book)
Exploring the Building Blocks of Science Book 1
Exploring the Building Blocks of Science Book 2
Exploring the Building Blocks of Science Book 3
Exploring the Building Blocks of Science Book 4
Exploring the Building Blocks of Science Book 5
Exploring the Building Blocks of Science Book 6
Exploring the Building Blocks of Science Book 7
Exploring the Building Blocks of Science Book 8

Focus Series unit study program — each title has a Student Textbook with accompanying Laboratory Notebook, Teacher's Manual, Lesson Plan, Study Notebook, Quizzes, and Graphics Package

Focus On Elementary Chemistry
Focus On Elementary Biology
Focus On Elementary Physics
Focus On Elementary Geology
Focus On Elementary Astronomy

Focus On Middle School Chemistry
Focus On Middle School Biology
Focus On Middle School Physics
Focus On Middle School Geology
Focus On Middle School Astronomy

Focus On High School Chemistry

Super Simple Science Experiments

21 Super Simple Chemistry Experiments
21 Super Simple Biology Experiments
21 Super Simple Physics Experiments
21 Super Simple Geology Experiments
21 Super Simple Astronomy Experiments
101 Super Simple Science Experiments

Note: A few titles may still be in production.

Gravitas Publications Inc.
www.gravitaspublications.com
www.realscience4kids.com

www.ingramcontent.com/pod-product-compliance
Lightning Source LLC
Chambersburg PA
CBHW082113210326
41599CB00033B/6682